A RETURN TO THE ICE AGE

The Ice Storms of 2000 and 2002

by

Oklahoma Living Magazine

THE
DONNING COMPANY
PUBLISHERS

A RETURN TO THE ICE AGE

The Ice Storms of 2000 and 2002

by

Oklahoma Living Magazine

Front cover photograph:
Courtesy Lenna White,
Western Farmers Electric Cooperative

Back cover photograph:
Courtesy Buddy Spradlin,
Lake Region Electric Cooperative

The Donning Company Publishers
184 Business Park Drive, Suite 206
Virginia Beach, VA 23462

Steve Mull, General Manager
Barbara B. Buchanan, Assistant General Manager
Ed Williams, National Sales Director
Jan Martin, Senior Copyeditor
Jeremy Gray, Graphic Designer
Scott Rule, Director of Marketing
Travis Gallup, Marketing Coordinator

Cataloging-in-Publication Data

A return to the ice age : the ice storms of 2002 and 2002 /
by Oklahoma Living Magazine
p. cm.
ISBN 1-57864-194-2
1. Ice storms—Oklahoma—History—20th century. 2. Ice storms—Oklahoma—History—21st century. 3. Electric power systems—Natural disaster effects—Oklahoma—History—20th century. 4. Electric power systems—Natural disaster effects—Oklahoma—History—21st century. I. Oklahoma Association of Electric Cooperatives.

QC926.44.O5 O38 2002
363.34'926'09766—dc21

2002035101

Printed in the United States of America

Table of Contents

Acknowledgments

We gratefully acknowledge the contributions made to this book by a number of groups and individuals. First and foremost, a special thank you to the member-owners of many electric cooperatives across Oklahoma. Since December 25, 2000, when the first of two devastating ice storms slammed into the state, members have shown tremendous patience and resilience. Their ability to cope under the most difficult of circumstances, and their extraordinarily positive attitude during a time of disaster, is reflective of the true pioneer spirit that still exists among Oklahomans.

This photograph, which shows the thickness of ice on a Cimarron Electric line compared with a coke can, has been used across the nation to demonstrate the severity of the storm. In some areas of the state, ice was four to six inches thick. (Courtesy Justin Glazier and Mike Burge, Cimarron Electric Cooperative)

To the scores of linemen, office, and right-of-way personnel, contractors, equipment operators, and vendors who worked tirelessly to keep supplies pouring in, which enabled thousands of miles of electric line to be rebuilt in mere days, we also say thanks. It is this kind of courageous and selfless dedication to "getting the lights back on" that permeates the rural electric industry.

There are several contributors that made this book possible. The staff of the Oklahoma Climatological Survey (OCS), in particular Derek Arndt, climatologist; Gary McManus, climatologist; Mark Shafer, senior staff climatologist/policy analyst; Kevin Kloesel, director of Outreach; and Dr. Ken Crawford, OCS director and state climatologist. Their willingness to provide scientific research along with their dedication to assist entities in accessing valuable weather data is simply immeasurable.

Additionally, the Oklahoma Department of Tourism, the Oklahoma Department of Agriculture, and the Oklahoma Department of Civil Emergency Management have made contributions that outline the impact the ice storms had on the entire state of Oklahoma.

We also owe a huge debt of gratitude to the Federal Emergency Management Agency (FEMA). Without their assistance, individuals, businesses, local governments, and certain nonprofit organizations would still be in a state of recovery.

And lastly, we recognize the directors, managers, and key staff members of all electric cooperatives affected by the Christmas 2000 and January 2002 ice storms. These are the men and women who directed the power restoration process, many of whom worked virtually around the clock. Their efforts should not go unnoticed, for they truly exemplified "the power of human connections."

Although it is hard not to notice the beauty of the ice as the sun glistens through the clouds, electric cooperatives spent weeks repairing the damage left behind by this beauty in disguise. (Courtesy Lenna White, Western Farmers Electric Cooperative)

Introduction

The words Oklahoma and weather have come to be synonymous. Each year the Sooner State is hit with a barrage of environmental extremes. It is not uncommon to feel the heat of a one-hundred-degree day in June that is followed just hours later by a cold front that drops temperatures into the mid- to upper-thirties. People who live within the confines of our seventy-seven counties know full well the power of the winds that come sweeping down the plain. Oft times these winds turn violent and spin up tornadoes that tear ugly paths of death and destruction across the countryside. Witness the May 3, 1999, tornado outbreak, when more than seventy twisters ripped helter-skelter across the state in a twenty-four-hour period.

Christmas 2000 was celebrated by many Oklahomans—in the dark. The ice that hit the state covered forty-six counties in the south and eastern quadrants of Oklahoma. (Courtesy Choctaw Electric Cooperative)

As damaging as tornadoes can be, they are, for the most part, found to touch down in largely uninhabited or sparsely populated areas of Oklahoma. Another common weather extreme is freezing rain, and depending upon the circumstances, this weather phenomenon can create its own special havoc—ice storms. In general terms, they happen somewhat frequently here; however, none compare with the significant ice storms that occurred on Christmas Day 2000 and in January 2002, two of the worst ice storms in Oklahoma history.

Both storms impacted different parts of the state: the Christmas 2000 ice event targeted forty-six counties largely in the south and eastern quadrants of Oklahoma, while the January 2002 storm chilled more than forty-five counties in the central, northern, and western parts of the state.

In the pages that follow, this book will attempt to explain why such ice storms occur and what atmospheric conditions combine to wreak utter devastation on trees and overhead electric lines. Photos from citizens and utility personnel will serve to document both the crystalline beauty of the ice and its destructive powers.

The Christmas 2000 ice storm struck heavily forested areas, which enhanced damage because of the volume of falling tree limbs. In all, sixty-five counties were declared federal disaster areas. (Courtesy Jack McLemore, Caddo Electric Cooperative)

While the paths of tornadoes are relatively hard to predict, certain factors can be analyzed to forecast, in general, an ice storm's footprint. Thanks in large part to the development of the Oklahoma Mesonet, real-time data is gathered every fifteen minutes at more than 115 stations around the state. This data is used by the Office of State Emergency Management, county emergency management officers, and, beginning in 1998, utility personnel from the Electric Cooperatives of Oklahoma.

Such entities are now better able to prepare in advance for these types of storms. While ice storms cannot be avoided, they can be dealt with in a much more effective and efficient manner thanks to today's technology that is available through the Oklahoma Mesonet.

It is our sincere hope that the following pages will serve to educate, enlighten, and enlarge each individual reader's understanding of ice storms and the people and businesses that they affect. One can only pray that there will never be another "return to the Ice Age" quite like the consecutive winter events that occurred in 2000 and 2002.

Chapter One
The Ice Storms Cometh

Editor's note: The following excerpts have been printed with the permission of the Oklahoma Climatological Survey and are taken from Publication ES 2002-01, which was authored by Gary McManus, staff climatologist. OAEC greatly appreciates the collaborative effort put forth by the Oklahoma Climatological Survey, the University of Oklahoma, and Oklahoma State University in developing and expanding the Oklahoma Mesonet.

Freezing rain is certainly not uncommon in Oklahoma. Almost every winter storm that visits the state leaves a bit of ice as a calling card along with the obligatory snow and sleet. Ice storms are a different story, however, especially those of the significant variety. They have struck throughout the state's recorded history with varying degrees of damage, and it is not unheard of to have two occur in rapid succession.

Thus, there was little shock when a year after a devastating ice storm struck portions of southeastern Oklahoma, the northwestern half of the state suffered the same fate with the January 2002 event. As meteorological conditions evolved, it was evident that the classical ingredients needed for a significant ice storm were coming to fruition.

When cooperatives put their Mutual Assistance Plan into action, other crews were brought in to help battle the destruction. (Courtesy Lenna White, Western Farmers Electric Cooperative)

The back-to-back ice storms caused nearly one-half billion dollars in damages to the state of Oklahoma. Much of the cost was attributed to debris removal because of the massive quantities of trees that were destroyed. (Courtesy Kiamichi Electric Cooperative)

What distinguishes one ice storm from another? Each seems to have a set of common factors. Variations revolve around several primary determinants: (1) ground temperatures at the time of the event; (2) levels of moisture above the surface for the formation of precipitation; (3) layers of warm air aloft, which allow for liquid precipitation; and (4) the amount of cold air at the surface so that rain will freeze on contact.

These differences were readily apparent during the two storms. The January 2002 ice storm pounded the state for little more than twenty-four hours, but its impact will be felt for quite some time. The storm left over a quarter million people without electricity and more than $200 million in damages, which prompted President George W. Bush to declare over half of the counties in Oklahoma federal disaster areas.

FEMA estimated the Christmas 2000 ice storm in southeastern Oklahoma reached $256.5 million in total damages. (Courtesy David McClure, People's Electric Cooperative)

Anatomy of the January 2002 Ice Storm

The pre-storm environment was decidedly warm with unseasonable temperatures in the days prior to the event. The high temperatures, lack of moisture, and high winds resulted in a major portion of Oklahoma being placed under a "red flag" fire alert.

Yard art may have been one of the few things that could appreciate the beauty of the ice. (Courtesy Loree DeCou)

The spring-like conditions allowed the relatively bare soil to begin storing heat. Soil temperatures in northwestern and central Oklahoma had risen into the mid-50s by the afternoon of January 28. The air began to grow colder by the afternoon, however, as a very shallow cold front crept in under the warm air. The cold front's southward progress halted in central Oklahoma near the El Reno and Oklahoma City metropolitan area.

Meanwhile, temperatures fell into the thirties in northern Oklahoma; however, south of the cold front, temperatures had once again soared into the seventies. The front's shallow nature was evident by the warmth lingering at the Mesonet's higher elevation stations in the western Oklahoma Panhandle near Kenton and Boise City. A push of more frigid air began overnight and into the afternoon hours of Tuesday, January 29, which dropped temperatures below freezing from Oklahoma City northwestward.

Precipitation began falling in earnest that evening, after elevated surfaces had cooled sufficiently. This allowed the liquid precipitation to freeze instantly upon contact and begin to accumulate.

The warm weather of the pre-storm environment kept ground temperatures unseasonably high, which prevented large-scale travel problems that are customary with other significant ice events, particularly the Christmas 2000 ice storm in southeastern Oklahoma and western Arkansas.

At one point during the 2002 ice storm, more than a quarter million people were without lights, heat, and electricity. (Courtesy Caddo Electric Cooperative)

Thank goodness for indoor plumbing! When the temperatures were well below zero, many Oklahomans were glad to have indoor bathrooms. (Courtesy Kristi Kelle)

A 6:00 A.M. atmospheric sounding from Norman on January 30 exhibited classic ice storm characteristics. A deep layer of warm, moist air above the surface was present, so the potential for liquid precipitation existed. In fact, with saturated air at a temperature close to fifty degrees at about five thousand feet above the surface, heavy rain was likely. Add to that a 1,330-foot-thick layer of below-freezing air close to the surface, and all the criteria for a "significant" icing event were realized.

Many farmers had to deal with the ice on a personal level. Water pumps were rendered useless, and bulk water supplies had to be delivered for cattle, horses, and other livestock. (Courtesy Jack Chapman, People's Electric Cooperative)

Linemen in the background replace poles as the stop sign shows the thickness of the ice that brought the line down. (Courtesy Kiamichi Electric Cooperative)

This stretch of fence along Chapman Farms south of Calvin shows the massive amounts of ice that hit parts of the state. (Courtesy Jack Chapman, People's Electric Cooperative)

A broken pole lays next to a frozen section of irrigation pipe in western Oklahoma. (Courtesy Caddo Electric Cooperative)

While the city of Norman did not receive any ice because of above-freezing surface temperatures, it was a different matter altogether just fifty miles to the northwest, where the surface temperature was below freezing for the duration of the event. As liquid precipitation fell through the freezing layer, it became "supercooled" and froze upon contact with elevated surfaces (i.e., tree limbs and electric power lines) at ground level.

The chart to the right shows a large amount of precipitation fell between January 29 and 30. Storm-total rainfall from the Mesonet shows the severity of the ice storm. The hardest hit areas, a broad swath from Elk City in the southwest to Pawhuska in the northeast, picked up as much as three inches of rain. This contributed to ice accumulations that on average were one to two inches thick across northern and west-central Oklahoma. Some local accumulations were between four and six inches. Oklahoma City shattered its previous record for January 30 rainfall with 2.13 inches, which broke the record of 1.34 inches set back in 1982.

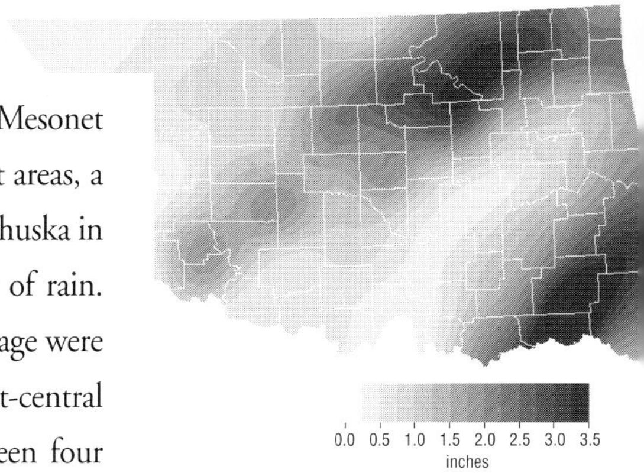

The weather's fickle nature was demonstrated by observing the fortunes of two closely located communities, El Reno and Norman. Meteorological conditions at the two central Oklahoma cities were very similar on the evening of January 29 and throughout the day on January 30 with one small exception—surface air temperatures. A small two-degree Fahrenheit difference in surface temperature resulted in vastly different fates for the two communities.

Both the Norman and El Reno Mesonet stations received approximately one and one-half inches of rain in the same twenty-four-hour period. However, at El Reno, where temperatures were just below freezing, the liquid precipitation froze instantly on elevated surfaces, which resulted in severe tree and property damage and downed power lines and utility poles. Two-thirds of this community of sixteen thousand was without electricity for several days. In contrast, temperatures in Norman hovered between thirty-two and thirty-three degrees for much of the same time period and only dropped below freezing for a few hours after

Icicles dangled from street signs, lampposts, and wires as they waited to hit the ground like spears. (Courtesy Michaleen Owens, Kay Electric Cooperative)

Fences bowed with the weight of up to six inches of ice. (Courtesy Buddy Spradlin, Lake Region Electric Cooperative)

the rain had lessened. Thus, Norman was spared the catastrophic damage suffered by its neighbor just forty-six miles to the north and west.

This same scenario played out across southwestern, central, and northeastern Oklahoma for the duration of the ice event. To the northwest of the freezing line, accumulating ice and devastation was the norm. Meanwhile, to the southeast a much-needed soaking rainfall was the predominant experience. The position of the boundary between sub- and above-freezing temperatures at noon on January 30 is indicative of the separation between damaged and undamaged locales (see map to the left). Tulsa area residents dodged the ice storm bullet by spending the majority of the twenty-four-hour storm period on the southeast side of the freezing boundary. If the surface temperatures in the state's second most populous city had dropped below thirty-two degrees, property damage and power outage figures would have increased dramatically.

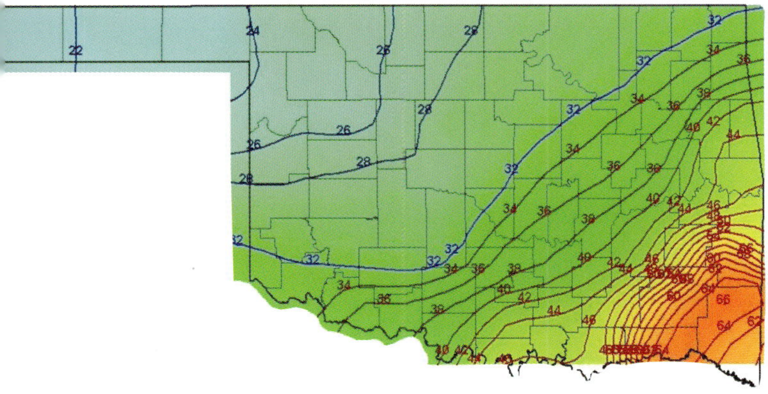

As the ice accumulated, tree limbs snapped and were left mangled and broken. (Courtesy Raymond Smith)

As the upper-level storm system that spawned the ice storm began to travel to the east, an expected second bout of heavy freezing rain did not develop. Indeed, even as precipitation ended late Wednesday evening, much of Oklahoma braced for another round of freezing mayhem from the skies, when the National Weather Service office in Norman issued heavy snow and ice storm warnings for the overnight hours. Instead, only intermittent bouts of freezing precipitation fell, as a dry incursion of air halted any widespread precipitation.

The 2000 Ice Storm

Caddo Electric Cooperative lost a total of 5,500 poles and 305.56 miles of line. (Courtesy Caddo Electric Cooperative)

The Christmas ice storm of 2000, arguably still the worst in Oklahoma history in terms of property damage, was especially severe in the southeastern quarter of the state. A mixture of sleet and freezing rain created havoc as travel was disrupted or, in some cases, halted completely late Christmas Day. Ice on tree limbs and power lines led to power outages in many areas. By one estimate, at least 170,000 homes and businesses, including 90 percent of the residents of McIntosh, Latimer, and Pittsburg Counties, lost power as a result of the storm. Power still had not been restored in some areas by the end of December 2000.

Several McCurtain County locations reported significant precipitation (liquid equivalent) over the three-day storm period, most notably Valliant, 5.56 inches; Broken Bow, 5.14 inches; and Carnasaw Tower, 4.33 inches. Damage to trees from ice buildup was significant throughout the forests of southeastern Oklahoma. Twenty-six deaths were

Nothing escaped the wrath of Mother Nature. While many Oklahomans waited for power to be restored, livestock faced the ice head-on in the fields, where hay was at a premium. (Courtesy Sondra Boykin, Western Farmers Electric Cooperative)

reported in connection with the storm system that brought ice to the region, although not all of those were directly attributed to the ice storm itself.

In terms of money, this ice storm takes top billing in state history with damages and other costs in excess of $250 million. In comparison to the January 2002 event, this ice storm affected less densely populated locales. The pre-storm environment, however, was decidedly colder than that of the January 2002 storm. The cold air was in place before the ice storm, which allowed soil temperatures to drop near or below freezing over much of the state. Thus, when the precipitation began, it immediately froze on the roads. This contributed to travel problems and hampered power line repair efforts.

The map below shows storm-total rainfall can be compared and contrasted with the ice storm of January 2002. Again, the main difference between the two storms was the significantly colder soil temperatures in southeastern Oklahoma during the days prior to the Christmas 2000 ice storm. Because soil temperatures were so much colder, ice and snow accumulations stayed on the ground for a much longer period of time and thus hampered debris removal and electric power restoration efforts.

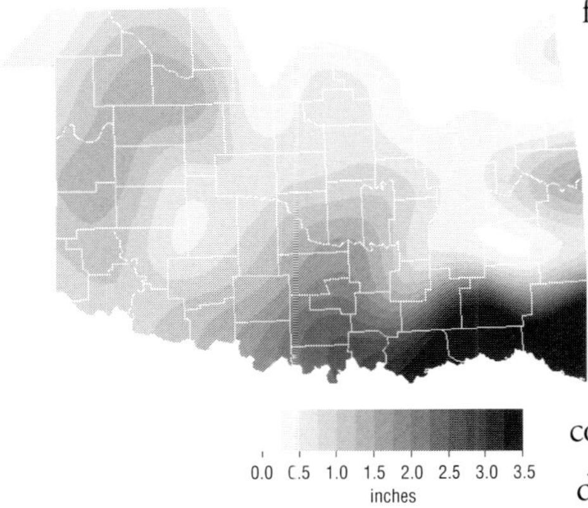

The Christmas 2000 ice storm struck heavily forested areas, which enhanced damage because of the sheer volume of falling tree limbs. In all, sixty-five counties in Oklahoma were declared federal disaster areas.

One of the 5,500 poles Caddo Electric Cooperative lost during the 2002 storm. (Courtesy Caddo Electric Cooperative)

Chapter Two

Aftermath and Damage Assessment

Below: As a result of the January 2002 ice storm, 38,030 distribution poles and 1,068 transmission structures were destroyed. (Courtesy Choctaw Electric Cooperative)

Collectively, the consecutive winter ice storms from Christmas 2000 and January 2002 caused in excess of $500 million in damages mainly to private and public properties, trees, forests, and electric systems. In particular, the state's electric power suppliers were hard-hit after the January 2002 storm. More than 255,000 people were without power, and some residential customers were without electricity for up to thirty-eight days.

A week after the weather system exited the state, more than thirty-nine thousand Oklahomans were still in the dark as utility companies worked around the clock to replace snapped poles and broken power lines. Enid, a city of forty-seven thousand people, was entirely without electricity for several days. High-voltage electric-transmission lines into the city from several different

directions lay in ruin on the ground, and structures succumbed to the massive weight of literally tons of ice. Initially, some electric suppliers feared it would take up to two months to rebuild lines, especially in rural areas, because damage was so extensive and the availability of materials and supplies was unknown.

Lines swayed to the ground under the weight of the ice. Central Rural Electric in Stillwater lost 2,200 poles and 122.22 miles of line. (Courtesy Kendra Stanek, Central Rural Electric Cooperative)

Tom Garrett, general manager of Cimarron Electric Cooperative, Kingfisher, reflected on the severity of the damage to his rural electric co-op's system. "No matter where you drove or where you looked, poles and line were down everywhere. I live about eight miles east of Kingfisher on Highway 33, and there are three-phase electric lines running alongside that highway for miles. There were only a few poles standing on the drive from my house into town. The devastation was absolutely incredible."

The ice storm's damage was so extensive that it took several days for utility engineers to fully comprehend its magnitude. Jon Joyce, manager of engineering for both Central Rural Electric Cooperative, Stillwater, and Indian Electric Cooperative, Cleveland, called initial assessment efforts "futile."

Cimarron Electric Cooperative in Kingfisher lost 15,500 poles and more than 861 miles of line during the 2002 storm. (Courtesy Mark Snowden, Cimarron Electric Cooperative)

"It's really difficult to precisely define the scope of damage to a utility when a storm of this size hits your system. In a matter of hours, our operations centers were inundated with phone calls from people without power. Unfortunately, you can't begin the repair process until you know what's broken. And believe me, there was an enormous amount of our electric system that was broken and laying on the ground," Joyce said.

In addition to downed lines and the cold, line crews had to contend with rural roads that were treacherous because of mud. (Courtesy Caddo Electric Cooperative)

In the early morning hours of January 30, engineers, operations managers, rights-of-way technicians, and power-line repair crews began the slow, deliberate process of estimating system losses. Beginning in far southwest Oklahoma in northern Greer County, major damage was reported on a line from near Granite and Sayre that extended northeastward to Enid and El Reno and then continued northward and eastward to Perry, Blackwell, Ponca City, Pawhuska, and Bartlesville. The storm continued into Kansas and Missouri, where freezing rain also took a toll on tree limbs and power lines.

This page: Along a stretch of Highway 15 west, not a single pole was standing. Transmission lines lay in shambles along the road. (Courtesy Shelley Berry, *Oklahoma Living* Magazine)

Twenty-two Oklahoma counties experienced major damage from the storm: Beckham, Blaine, Caddo, Canadian, Custer, Dewey, Garfield, Grant, Greer, Kay, Kingfisher, Kiowa, Logan, Major, Noble, Oklahoma, Osage, Payne, Pawnee, Roger Mills, Washington, and Washita. A Federal Emergency Management Agency (FEMA) disaster declaration was issued on February 1, 2002. It included twenty-three additional counties that sustained varying degrees of destruction: Alfalfa, Beaver, Cimarron, Cleveland, Comanche, Creek, Ellis, Garvin, Grady, Harmon, Harper, Jackson, Lincoln, McClain, Nowata, Pottawatomie, Rodgers, Stephens, Texas, Tillman, Tulsa, Woods, and Woodward.

The damage to electric cooperative systems from the December 2000 storm was estimated at $38.5 million. (Courtesy Mark Snowden, Cimarron Electric Cooperative)

Across those counties scores of communities served by Oklahoma Gas & Electric Company (OGE), American Electric Power (formerly Public Service Company of Oklahoma), Oklahoma Municipal Power Authority (OMPA), and the electric cooperatives of Oklahoma were plunged into darkness within a twenty-four-hour period. At one time more than a quarter-million residents (approximately 8 percent of the state's population) were without lights, heat, and electric power.

Twelve electric-distribution cooperatives and two generation-transmission cooperatives sustained considerable losses. Six of those systems—Caddo Electric Cooperative, Binger; Central Rural Electric Cooperative, Stillwater; Cimarron Electric Cooperative, Kingfisher; Kay Electric Cooperative, Blackwell; Kiwash Electric Cooperative, Cordell; and Western Farmers Electric Cooperative, Anadarko—were especially hard-hit. The chart on the following page shows estimated utility pole losses as a result of the January 29–30 storm.

Name of Electric Cooperative	# of Poles Destroyed	Miles of Line
Alfalfa Electric Co-op, Cherokee	100	5.56
Caddo Electric Co-op, Binger	5,500	305.56
Central Rural Electric Co-op, Stillwater	2,200	122.22
Cimarron Electric Co-op, Kingfisher	15,500	861.11
Harmon Electric Association, Hollis	110	6.11
Indian Electric Co-op, Cleveland	450	25.00
KAMO Power*, Vinita	50	5*
Kay Electric Co-op, Blackwell	9,000	500.00
Kiwash Electric Co-op, Cordell	4,250	236.11
Northfork Electric Co-op, Sayre	615	34.17
Northwestern Electric Co-op, Woodward	5	<1
Oklahoma Electric Co-op, Norman	100	5.56
Southwest Rural Electric Association, Tipton	100	5.56
Verdigris Valley Electric Co-op, Collinsville	100	5.00
Western Farmers Electric Co-op*, Anadarko	1,036	104*

* generation-transmission co-op
Totals: 38,030 distribution poles; 1,086 transmission structures (equals 2,113 miles of distribution line; 109 miles of transmission line).

The trees that surrounded Steve and Jo Randall's home in Porum were left in shambles after the 2002 ice storm. (Courtesy Steve and Jo Randall)

Right: Some of the most remote areas of the state were the hardest hit because of the abundance of trees surrounding the lines. (Courtesy Rural Electric Cooperative)

Below: Twenty-two counties experienced major damage during the 2002 storm. Twenty-three additional counties sustained varying degrees of destruction. (Courtesy Jack McLemore, Caddo Electric Cooperative)

Below: Kay Electric Cooperative in Blackwell lost over 500 miles of line during the January 2002 ice storm. (Courtesy Shelley Berry, *Oklahoma Living* Magazine)

Above: The January 2002 ice storm caused an estimated $130 million in damages to electric cooperatives systems, and FEMA assistance now totals nearly $98 million. (Courtesy Lenna White, Western Farmers Electric Cooperative)

Right: According to the Oklahoma Department of Agriculture's Forestry Division, thousands of trees were lost or damaged during the ice storms that hit the state. Many of the trees still show the damage today. (Courtesy Rocki and Karen Hunsucker)

Left: The January 2002 ice storm had seven deaths attributed to it, and twenty-six lives were lost during the Christmas 2000 ice storm. (Courtesy Opal Bray)

Comparatively, seven electric cooperatives in southeastern Oklahoma sustained collective losses of sixty-five hundred poles during the Christmas 2000 ice storm. Approximately twenty miles of transmission line went down as well. Damage to rural electric infrastructures is estimated at $38.5 million, and FEMA reimbursements to date approximate $29 million. The January 2002 ice storm has inflicted an estimated $130 million in damages to electric cooperative systems to date, and FEMA assistance now totals nearly $98 million.

Above: More than 255,000 people were without power during the January 2002 ice storm. (Courtesy Caddo Electric Cooperative)

In terms of total costs for debris removal, emergency protective measures, and damage to roads, bridges, water control facilities, public buildings, public utilities, recreational, and other facilities, FEMA estimates the January 2002 storm will exceed $203.75 million with eligible reimbursements to exceed $153 million. By contrast, the Christmas 2000 ice event in southeastern Oklahoma reached $256.5 million in total damages incurred with $192.36 million in FEMA reimbursements to eligible participants.

Right: The ice storm damage hit electric cooperatives so hard that it took several days for utility engineers to realize the magnitude of the damage. (Courtesy Kendra Stanek, Central Rural Electric Cooperative)

A week after the weather system left the state, more than 39,000 Oklahomans were still in the dark as line crews worked around the clock to replace snapped poles and broken power lines. (Courtesy Rural Electric Cooperative)

The primary difference in costs between the two storms revolves around debris removal. FEMA funded debris removal at 100 percent for the Christmas 2000 ice storm because of the massive quantities of trees that were destroyed and littered county and state roadways. More than $52 million was spent on debris removal alone.

The two ice storms that hit Oklahoma during Christmas 2000 and January 2002 caused more than $500 million in damages. (Courtesy Steve and Jo Randall)

Above: Ice combined with heavy winds caused poles to snap like pencils. (Courtesy Caddo Electric Cooperative)

Left: A Federal Emergency Management Agency disaster declaration was issued on February 1, 2002. (Courtesy Kendra Stanek, Central Rural Electric Cooperative)

Opposite page, above: This is a photograph of Robert and Billie Brown's home before the 2002 ice storm. The photograph below shows the damage to the Browns' tree after the 2002 ice storm. (Courtesy Robert and Billie Brown, People's Electric Cooperative)

Collectively, these back-to-back ice events caused nearly a half-billion dollars in damages. The single most devastating weather event in state history was the May 3, 1999, tornado outbreak, which caused an estimated $1.26 billion in property and infrastructure loss. Forty-four people were killed during that tornado event, which swept through southern, western, central, and north-central Oklahoma. The January 2002 ice storm had seven deaths attributed to it, and twenty-six lives were lost in the Christmas 2000 ice storm.

In both events there were additional secondary effects: fences were destroyed and thus allowed cattle out, which disrupted traffic; many water treatment plants were temporarily without power, so boil orders had to be authorized by the Oklahoma State Department of Health in affected areas; and water pumps were rendered useless and thus necessitated bulk water deliveries for cattle, horses, and other livestock.

Chapter Three

The Cavalry Responds

Because National Weather Service forecasters had provided some warning to local and state officials nearly forty-eight hours in advance, road crews and utility repair technicians and contractors were on alert. Sand and salt trucks were readied as early as Monday, January 28. Electric cooperatives began contacting sister co-ops that were not projected to be in the path of the storm and put them on notice that additional manpower might be needed. In addition, material vendors were notified about the impending storm.

Crews from cooperatives in southeastern Oklahoma, who had received help during the Christmas 2000 ice storm, were some of the first to volunteer during the 2002 ice storm. (Courtesy Juli Orme, Cookson Hills Electric Cooperative)

Since cooperative line crews are trained to the same safety and construction standards, they were able to work on lines anywhere in the state. (Courtesy Rural Electric Cooperative)

Bulldozers were brought in to assist trucks that became stuck as lines were repaired. (Courtesy Juli Orme, Cookson Hills Electric Cooperative)

While there is little that can be done to prevent ice damage, much preparation can be done by utilities in advance of such disasters. Emergency response plans are evaluated; mutual aid agreements and disaster assistance plans are reviewed and upgraded; and additional crews are put on standby—ready, willing, and able—to begin repair work in a moment's notice.

Such was the case on Wednesday, January 30, 2002. Earlier in the week, Kenny Guffey, director of Safety and Loss Control for the Oklahoma Association of Electric Cooperatives (OAEC), Oklahoma City, was on the phone contacting the association's thirty member cooperatives and evaluating resources at their disposal. Sid Sperry, OAEC's director of Public Relations, Communications, and Research, had spent Monday, January 28, reviewing weather forecasts from the National Weather Service and the National Severe Storms Laboratory in Norman. The electric cooperatives of Oklahoma partnered with the Oklahoma Climatological Survey (OCS) in 1998 to provide Oklahoma's rural electric cooperatives with real-time radar, Mesonet, and other meteorological data. The information derived is used to reduce consumer costs, to keep power line technicians safe when working in the field, and to anticipate outages.

"Weather data was placed into the hands of dispatchers from our member systems in early 1999," said Sperry. "We coordinated with officials from OCS, and a series of three instructional workshops were held to train cooperative personnel in the art of interpreting and applying weather data to electrical operations." Sperry added, "Proper use of the data allows each cooperative to operate more safely and reliably during times when delivering electricity might be compromised, like during an ice storm."

Severe weather events, such as thunderstorms, ice storms, and windstorms, can cause widespread outages to a utility's transmission and distribution system. By using data made available from the Mesonet and from OCS forecast information, electric cooperatives can better plan for outages and actually reduce outage times to end-use consumers.

Vehicles that carried supplies to the line crews had to be constantly refilled and returned to the field. (Courtesy Caddo Electric Cooperative)

Left: Co-op employees showed the same pioneer spirit that their ancestors did when they brought electricity to the countryside some sixty years ago. (Courtesy Kendra Stanek, Central Rural Electric Cooperative)

Right: In addition to in-state crews, crews also were called in from eleven other states. (Courtesy Kendra Stanek, Central Rural Electric Cooperative)

Below: Line crews filled country roads as they worked diligently to restore power. (Courtesy Rural Electric Cooperative)

Forecasts on Sunday, January 27, and Monday, January 28, prompted Sperry to send out an ice storm alert on Monday afternoon, which was followed by an ice storm warning on Tuesday morning. The warning listed counties where freezing rain and ice accumulations of more than one-half inch were anticipated.

"The forecasters at the National Weather Service and OCS did a tremendous job in predicting the path and footprint of this storm," said Sperry. "For the most part, their forecasts were right on target. It was just a matter of knowing how much freezing rain might fall, and that was difficult to estimate."

As the rain continued late into Tuesday night, Guffey stayed in contact with cooperative operations managers and line superintendents to assess early damage reports and to activate the cooperatives' Mutual Assistance Plan.

"We started calling co-op managers whose systems were located outside the footprint of the storm first, then began the process of contacting out-of-state co-ops, and then contract repair crews and contract right-of-way crews," Guffey said. "By the time it was all said and done, we had crews from eleven states working to try and get power back on. The response was overwhelming."

These trucks with new poles were headed to Cimarron Electric Cooperative to replace 15,500 poles that were downed during the 2002 ice storm. (Courtesy Mark Snowden, Cimarron Electric Cooperative)

Line workers from the unaffected systems in Oklahoma were quickly dispatched to those systems in need. Crews from cooperatives in southeastern Oklahoma, who had been recipients of assistance from sister co-ops in other parts of the state during the Christmas 2000 ice storm, were some of the first to volunteer.

Kay Electric set up temporary storage areas for large piles of new poles. (Courtesy Kay Electric Cooperative)

"Our line crews wanted to go and help out, because they had been on the receiving end of such aid only a year earlier," said Kendall Beck, general manager of Cookson Hills Electric Cooperative, Stigler. "That's why we call it a 'Mutual Aid Plan'; it is an action plan that matches repair crews, equipment, and material to the needs of those systems that are temporarily disabled."

Joe Harris, general manager of Kay Electric Cooperative, Blackwell, was on the receiving end of such assistance.

"We are a small cooperative with only about fifty-eight hundred meters and thirty-five full-time employees. When you have a disastrous event of this magnitude, it's important to know where your resources are going to come from. We can't say enough good things about the crews from other co-ops that came in to help rebuild our system. They did an incredible job, and, most importantly, did it safely," Harris said.

Line technicians and equipment operators from Alfalfa Electric Cooperative, Cherokee; Arkansas Valley Electric Cooperative, Ozark, Arkansas; Canadian Valley Electric Cooperative, Seminole; Choctaw Electric Cooperative,

Many members of Central Rural Electric Cooperative in Stillwater were dealt a double blow from Mother Nature with the May 3, 1999, tornadoes and then the January 2002 ice storm. (Courtesy Kendra Stanek, Central Rural Electric Cooperative)

Crews and office workers were at Kay Electric Cooperative twenty-four hours a day as they worked to get the lights back on for their members. (Courtesy Kay Electric Cooperative)

Many rural areas had to deal with destroyed fences, loose livestock, and useless water pumps until line crews could reach their farms. (Courtesy Caddo Electric Cooperative)

Because the National Weather Service had forecast the storm, electric cooperatives were able to start contacting sister co-ops that were not in the projected path of the storm for standby assistance. (Courtesy Buddy Spradlin, Lake Region Electric Cooperative)

The Mutual Assistance Plan proved how beneficial it was during the two back-to-back ice storms. (Courtesy Jack McLemore, Caddo Electric Cooperative)

The early morning sunrise shed light on line crews that had worked through the night. (Courtesy Buddy Spradlin, Lake Region Electric Cooperative)

Trucks carrying new poles to Central Rural Electric Cooperative in Stillwater could be seen for miles. (Courtesy Kendra Stanek, Central Rural Electric Cooperative)

Hugo; Cookson Hills Electric Cooperative, Stigler; Cotton Electric Cooperative, Walters; East Central Oklahoma Electric Cooperative, Okmulgee; KAMO Power, Vinita; Kiamichi Electric Cooperative, Wilburton; Lake Region Electric Cooperative, Hulbert; Northeast Oklahoma Electric Cooperative, Vinita; Northwestern Electric Cooperative, Woodward; Oklahoma Electric Cooperative, Norman; Ozarks Electric Cooperative, Fayetteville, Arkansas; People's Electric Cooperative, Ada; Red River Valley Rural Electric Association, Marietta; Rural Electric Cooperative, Lindsay; Southeastern Electric Cooperative, Durant; Southwest Rural Electric Association, Tipton; Tri-County Electric Cooperative, Hooker; and Verdigris Valley Electric Cooperative, Collinsville, all went to the aid of their stricken brethren.

The response from out-of-state co-ops and contractors also was impressive. Crews from Alabama, Arizona, Arkansas, Florida, Kansas, Kentucky, Louisiana, Missouri, Nebraska, New Mexico, and Texas raced to assist in the recovery effort.

Office personnel from other in-state cooperatives also were an integral part of the Mutual Assistance Plan. Jennifer McCall, director of Member, Education, and Youth Services for OAEC, coordinated the effort to supply affected co-ops with staff members who had member relations, customer service, engineering, and accounting skills. "It was overwhelming the way people from across the state pitched in to help their neighbor cooperatives in a time of need," said McCall. "We were inundated with volunteers and actually had to turn some people away. It really makes a person feel good about the organizations we work for."

Perhaps the most amazing response of all came from co-op members, some of whom were without electricity in their own homes and businesses for thirty-eight days.

"Our member-owners were simply unbelievable," said Dennis Krueger, general manager of Kiwash Electric Cooperative, Cordell. "They had been through so much in the last two years—windstorms, the October 2001 tornado, and then this ice storm—it was enough to test their spirit of endurance and perseverance."

Yet, that same pioneer spirit that brought electricity to the countryside more than sixty years before did in fact endure the twin ice storms of 2000 and 2002. Local church groups and extension clubs helped make daily meals for hungry line workers and equipment operators. Residents who were still without power drove for miles to restaurants or other facilities that did have electricity in order to prepare sandwiches or boxed lunches. Laundry services were provided in many instances.

The parking lot at Central Rural Electric Cooperative was a flurry of activity as line crews came to assist. (Courtesy Kendra Stanek, Central Rural Electric Cooperative)

Left: The ice storm of 2000 had many line crews working in some of the most difficult situations. (Courtesy Choctaw Electric Cooperative)

Right: Line crews worked around the clock to restore power. Many saw the beauty of an Oklahoma sunset from the top of a bucket truck. (Courtesy Sondra Boykin, Western Farmers Electric Cooperative)

"People were so understanding," said Don Crabbe, general manager of Central Rural Electric Cooperative, which is headquartered in Stillwater. "Many of our members had been through a double blow. They experienced the May 3, 1999, tornadoes, and then some of those same customers got hit with the January 2002 ice storm. From their perspective it would be easy to be discouraged, but they weren't. Instead, they came to us with comments like, 'What can I do to help?' which was really inspirational for all of our co-op employees to hear."

Below: Line crews had to fight brush in some rural areas in order to restore power. (Courtesy Caddo Electric Cooperative)

Said Bob Thomasson, chief executive officer of Caddo Electric Cooperative, Binger, "Our member-owners realized very quickly that this was an absolutely devastating storm. They understood that our cooperative employees and board of directors were doing everything possible to rebuild the electric system in order to get lights back on as soon as possible. They didn't complain one bit, which is both admirable and appreciated."

"It was really inspiring to see the determination of everyone involved in efforts to restore power," stated Don Crabbe, general manager of Central Rural Electric Cooperative. (Courtesy Kendra Stanek, Central Rural Electric Cooperative)

Crabbe, who is currently serving as the Oklahoma Association of Electric Cooperatives' president, said the dedication and response of both members and employees was inspiring. "It was really inspiring to see the deter-

Line crews worked in unbearable cold to rebuild poles and lines in some of the most remote areas of Oklahoma. (Courtesy Rural Electric Cooperative)

mination of everyone involved in efforts to restore power," said Crabbe. "All of our employees realized what a hardship it was for people to be without electricity, especially for weeks at a time. But our employees and our members acted as a team—it was a Herculean effort."

Echoing those sentiments were Johnie Grimes and Stephen Bryan, managers of Choctaw Electric Cooperative and Kiamichi Electric Cooperative, respectively, whose co-ops and memberships endured the Christmas 2000 ice storm. "We had some complaints, obviously," said Grimes. "But by far and away, the vast majority of our folks were very understanding and complimentary of our efforts to restore their power." Added Bryan, "People have come to take electricity for granted. It's supposed to be on every second of every day, seven days a week, every week of the year. But for the most part, our members understood that we were making every effort to get lines put back up in the air in a safe but efficient manner. No one jumps for joy when the power goes off, but in an extreme circumstance like the ice storm, people understand why."

Many linemen worked to restore power to co-op members, even when their own families were without. (Courtesy Rural Electric Cooperative)

Chapter Four

Words of Appreciation

In the days and weeks that followed the twin ice storms of 2000 and 2002, letters poured in to co-op offices from across the state. Here are a few examples of the words of encouragement that members conveyed to linemen and repair crews who labored mightily to rebuild hundreds of miles of electric line. They did this in less time than it took to rebuild the damaged I-40 bridge near Webbers Falls, two sections of which were destroyed on June 30, 2002, when a barge that was traveling the Arkansas River crashed into support piers that elevated the bridge.

During this winter's ice storm, our conversation often turned to the sincere appreciation we felt for the many people working in abhorrent conditions to restore our power.

I came home from work one day to find several trucks and linemen in a basket truck working just across the road from my house. I also had a snowman in my front yard, apparently the work of my 12-year-old twin boys just home from school. I cringed and wondered if my boys had been nuisances to the linemen. After the linemen left, I walked out onto my front yard to see the snowman. A neighbor stopped on the road and questioningly looked at my front slope. He said, "Thanks?" When he figured out that I was puzzled, he explained the large letters in the snow that were not visible from my house or my driveway. My sons had offered their gratitude to the linemen by carving a large "THANKS" in the snow that was visible from the basket. I don't know when I have ever been a prouder mom!

My sons said it better, but thank you to everyone involved.

Jill Humphrey
Kiwash Electric Cooperative

Line crews and office personnel were fed up to three meals a day and even had their laundry done by volunteers. (Courtesy Kay Electric Cooperative)

From the Wife of a Lineman

I'm proud to say—my husband's a lineman. The work hours for a lineman are very unpredictable. In addition to a regular work week, it's not uncommon for the phone to ring all hours of the night or early morning for a power outage at any given place. This means getting out of his warm bed and going into the night to fix whatever trouble that's caused the electricity to go off. If it's a thunderstorm, the work could be an all-nighter, then continuing to work the next day as if he had gotten rest. Linemen are "EVER-READY"—know it and believe it, because it's so.

I've read several articles, heard several opinions and have seen much damage as a result of the recent ice storm. Many families suffered loss in the ice storm of 2002. Even now, the ice is gone, but by looking over the countryside, it's plain to see something bad has happened.

The occupation of a lineman is one that not a lot of people want. It's a job that requires a lot of guts and little glory. A lineman risks his life each day to bring us lights. Sounds crazy—but it's true. You see, every pole that dots the landscape has been climbed by a lineman at one time or another. When you drive around and see miles of poles and line, a service pole or yard light, know that some lineman had to climb that stick. After strapping on his belt and hooks (which weigh about as much as a saddle), he climbs the pole. That pole could be 30 feet, maybe 40 feet or more. He does his work in the air after safetying off.

Working with electricity is like being around a rattlesnake—you've got to watch that it doesn't get you. Electricity commands respect. You've got to know what to do when you get to the top of the pole, whether in a bucket or safetied to the pole. Even then, sometimes accidents happen.

Electricity is a convenience we all enjoy having. We pay for it monthly, and yes we deserve to get what we pay for. And then there comes a problem called ice. First it's drizzle, then rain, then freezes and it's ice. The ice grows on the trees and lines until finally they can take no more—and they snap. There were thousands everywhere. You remember, they sounded like shotguns going off all over. While all this was going on, a lot of us were in the dark, literally. But the linemen were out in the drizzling, freezing rain working non-stop. Although they were cold and wet, they were working for us, to bring warmth and light back to our homes. Now, not only do they deal with iced poles and ice coated limbs covering lines and ice covered roads, but they must do this with very little rest. And this goes on for days—weeks. And they don't stop. In the worst weather, horrible working conditions and a never ending work load, they keep going.

You might say—well, that's their job—that's what they get paid for. To that I'd say—YES—it is their job, but at what cost should a person be asked to risk their life on a daily basis. What job description says, "Willing to risk it all . . . for you"?

No, linemen aren't gods, but can you remember how happy you were when your lights came back on, and stayed? Well, believe me when I say, they were even more happy than you! These men I'm describing are sons, brothers, husbands, daddies, grandpas or a loved one to one of us. Not only is my husband a lineman, but my brother and brother-in-law are linemen, too.

It took many hard working dedicated people working hand in hand to get us through this. And I'll say, not all of them were linemen. Thank you all so very much for your

The members of the Hinton Missionary Baptist Church showed their thanks to Caddo Electric Cooperative on the church's message board. (Courtesy Shelley Berry, *Oklahoma Living* Magazine) *many hours of work. We appreciate you so very much. For the people who were there with a hot cup of coffee—a meal—a chainsaw—a smile—a word of praise, I say thank*

you for helping to get through this disaster. For the in-state workers and the out-of-state workers who gave of their time so unselfishly . . . Thank You. Lives were lost, precious lives that cannot be replaced. They and their families have paid the ultimate sacrifice and will never be forgotten. The ice storm affected us all in many different ways.

Now the days roll on and normal life as we know it resumes. Workers are still working each day to keep our lives moving smoothly along.

To all of you we say "Thank You"—our hats are off to you and with a firm handshake, we say—you are appreciated! Job well done.

Debra Kerr
People's Electric Cooperative
2-14-01

The response to the ice storm from co-op members was outstanding. Students and parents at Frink Chambers Elementary School sent a goodie-bag to line crews at Kiamichi Electric Cooperative. (Courtesy Cathey Heddlesten, Kiamichi Electric Cooperative)

Hail Kay Electric

I've heard stories from long ago,
Of how my grandparents walked to school in the snow.
I've wondered what stories my grandchildren will hear,
Let me tell you what happened early this year.
It was the last of January, rain started to freeze.
Just one day out of school, Oh, P-l-e-a-s-e.
Be careful what you wish for, it might come to be
The most dreadful experience you ever did see.
Power lines sagged and poles snapped to the ground,
Nowhere in the county electricity was found.
We took baths and ate by candle light for days,
The highlight of one evening was playing charades.
Day two, day three, day four, and day five,
With no T.V. how can we survive?
Neighbors came for water, and friends came for baths
For we were the few with water and gas!
Day six, day seven, and finally day eight,
A generator came. Oh, my! It was great!
Two more days, we made it through
The terrible ice storm of 2002.

By Karli Hostetler, who just finished fifth grade at Billings Public School. Karli is eleven years old, and her parents are members of Kay Electric Cooperative.

Dear employees of Caddo Electric Cooperative,

I want to thank each and every one for all that you did during the ice storm. I want you to know that my heart went out to those that were out in the cold. I said prayers for all. I stayed in the shelter here at the Field House and was thankful that I did have a place to stay, warm food, hot coffee or tea. I know you were out in the cold with no relief in sight. How could I complain when all was being done that any human could do? I'm proud to be an American, but double proud that we have companies in America like Caddo Electric Cooperative.

Thank you again each and every one.

God bless all,
Mona Whiten

Old Glory stands high as people from across Oklahoma banded together to restore lights to the countryside. Their efforts truly exemplified "the power of human connections." (Courtesy Lenna White, Western Farmers Electric Cooperative)

Children from a nearby elementary school sent goodie-bags with words of encouragement to employees of Cimarron Electric Cooperative. (Courtesy Shelley Berry, *Oklahoma Living* Magazine)

Epilogue

In the days and weeks since the January 2002 ice storm, thousands of miles of electric line have been rebuilt. More than eleven hundred transmission structures and more than thirty-eight thousand electric-distribution poles have been replaced within rural electric co-op systems alone. The work to ensure a safe, reliable, and code-compliant system will continue for years to come. As the work continues, expressions of gratitude continue as well.

Special thanks go to every person from the Oklahoma Department of Civil Emergency Management who was involved in recovery efforts from both the Christmas 2000 and January 2002 ice storms. These dedicated and tireless individuals spent long hours away from their homes and families in order to provide emergency services to the citizens of Oklahoma, and we thank them for all the work that they do.

Words of appreciation also are directed to the officials with FEMA, who came from many regions of the country to assist Oklahomans in the recovery effort. Without FEMA assistance, direction, and funding, the restoration process would have been insurmountable. A debt of gratitude and countless words of appreciation go out to FEMA Director Joe Allbaugh, who is an Oklahoma native. His tour of western and northern parts of the state led him to understand the utter destruction that was wrought upon his fellow Oklahomans. He looked, he listened, and he cared about the home folks. Thank you, Director Allbaugh.

Another huge expression of appreciation goes out to Senator Jim Inhofe and Congressman Frank Lucas. They personally intervened on behalf of co-op and municipal electric consumers across the state and made staff members readily accessible to assist in any way they could. To both Senator Inhofe and Congressman Lucas and their entire staffs, we say thank you for your guidance and assistance.

Oklahoma Governor Frank Keating also deserves praise for his actions following the twin ice storms. Governor Keating and his staff have been most helpful in working with the electric cooperatives of Oklahoma during times of weather-related disasters. We deeply appreciate the efforts put forth by the governor and his staff.

To the contract repair crews who came from many states and who sacrificed time away from their families, we are most appreciative of your work and your dedication to getting the job done. We could not have accomplished the near miraculous work in the time frame in which it was completed without your help—thanks.

To each and every rural electric cooperative employee, whether from the state of Oklahoma or from surrounding states, we have been humbled once again by "the power of human connections." You know your duties, you know how to be part of an incredible team, and most of all, you have a sense of dedication to the membership that is unequalled in the business community. Words cannot express the amount of gratitude that is owed to each and every one of you.

Lastly, to the member-owners of each affected electric cooperative, we would like to simply say, "You're the best." Your patience, understanding, and willingness to help and to lend a hand is something that will not soon be forgotten. Thank you for being our owners, and thank you for being so kind and considerate during the most challenging of times. We know you can readily identify with the following quote from a farmer giving witness in a rural Tennessee church during the early 1940s: "Brothers and sisters, I want to tell you this. The greatest thing on earth is to have the love of God in your heart, and the next greatest thing is to have electricity in your house."

Preach on, brother, preach on!

Sidney K. Sperry

Sidney K. Sperry
Director of Public Relations
Oklahoma Association of Electric Cooperatives, Inc.

Shelley Berry

Shelley Berry
Executive Editor
Oklahoma Living Magazine